# **QUIRK** The curious case of a small island

You don't have to travel far in the world to realise there's nowhere remotely like this country.

Whether it's the peculiar place names, our eccentric events or our curious collectors, the truth is this modest island has more idiosyncrasies per square inch than any other nation.

Which is where this little volume comes in. It gives a glimpse into the curious richness this island possesses. It cocks an eye at the abundance of quirk all around us.

Read it. You just couldn't make it up.

QUIRK Published in the UK by The Number (UK) Ltd, Winchester House, 259 – 269 Old Marylebone Road, London NW1 5RA
www.thenumber118118.com

Published in 2004
Author Emily Spencer

ISBN 0-9547596-0-5

Ssshhhhhhh...

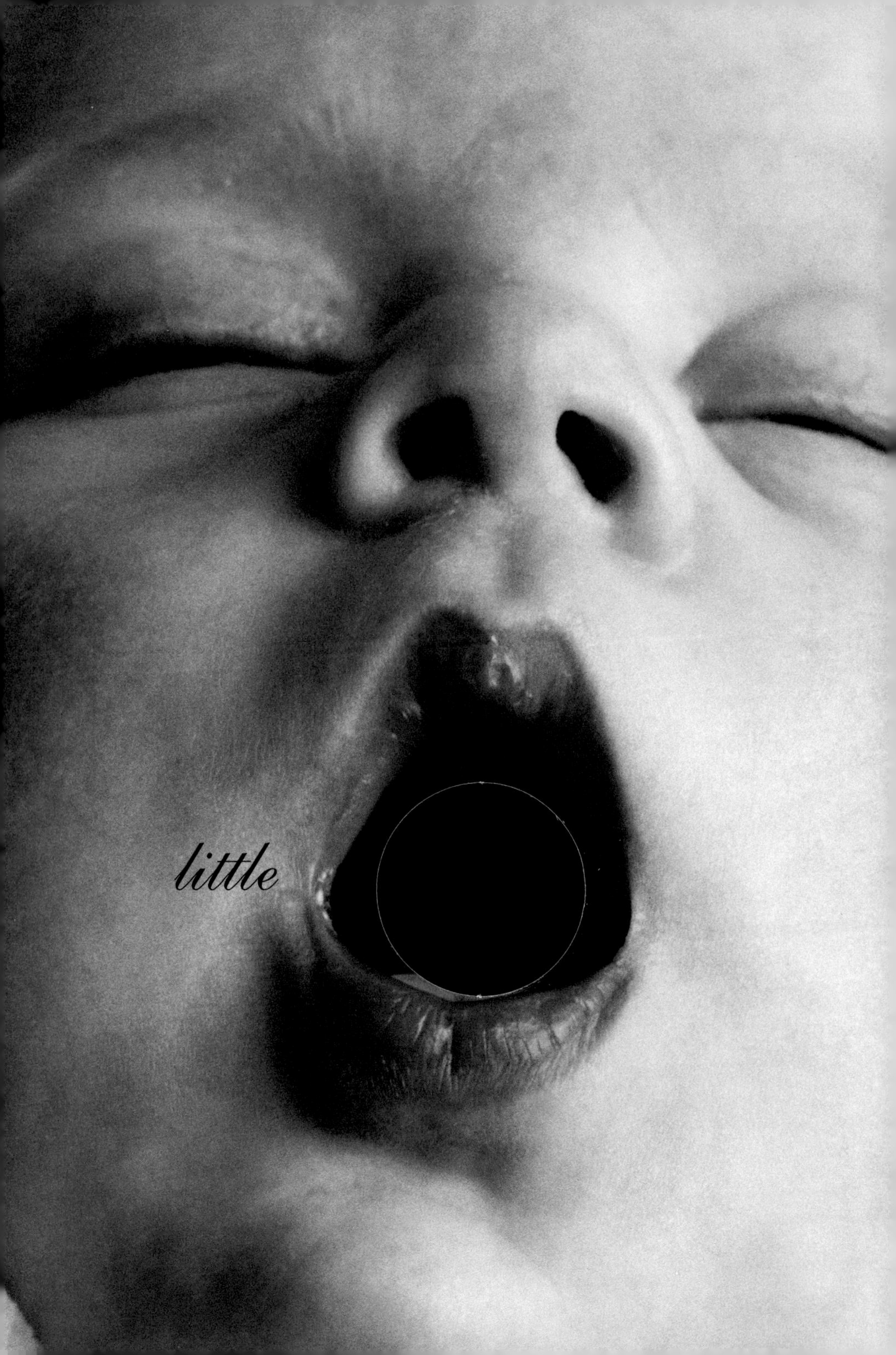
little

G
RAISING POULTRY
THE MODERN WAY

Little Snoring and Great Snoring rest side by side in tranquil Norfolk countryside. The bigger of the two is the Little, obviously. The Manor House has Great Snoring rooms where you can lay your sleepy head. Apparently, it's very quiet and peaceful.

*snoring*

Ogle

Ogle is a hamlet in Northumberland, but the nearest optician is about seven miles away. The first known Ogle or Hoggell or Oggel or Hoggal, after whom the place is named, was Humphrey, who was granted the lands during the reign of William the Conqueror. The Ogle family motto is “Prenez en Gré” or “Accept in Gratitude”.

C
UT
OUT
ANDT
RYTHEM

# The Handlebar Club of Great Britain

The fantashtic heroes of the Handlebar Club, the world's oldest moustache society, meet once a month at the Windsor Castle pub on Crawford Place, London. They represent Britain on the world's whiskered stage and sport a variety of guises, from the Fu Manchu, Dali, Imperial, and the Wild West. Ted Sedman, the Club's president, is the proud owner of the biggest Handlebar in the world, a Guinness World Record not to be sniffed at.

Cultivating this facial furniture is no mean feat, explains Michael 'Atters' Attree (above), Handlebar member and founder of the illustrious Ministry of the Moustache. His own upper lip requires 25 minutes' attention a day, and he has to sleep with a net – or snood – across his face. Pretenders beware, and beards, heaven forbid: his ministry offers sanctuary to the moustache wearer only: "The 'goatie' (waxed or otherwise) may revel elsewhere, in its own mediocrity."

Series 0.1
Dali
Series 0.12
Fu Manchu
Series 0.9
Attlee
Series 0.7
Pierre
Series 0.8
Wild Bill
Series 0.3
Garnett
Series 118
The Boys

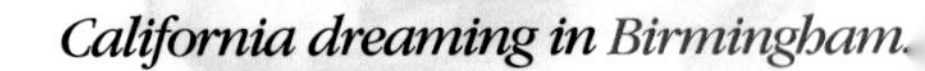
*California dreaming in* Birmingham.

Some places just aren't where they make out. You'd be forgiven for thinking that New York was in New York, and California was in America, and America was next to Canada, and Quebec was in Canada, next to America, and New Zealand, on the other side of the world. And Normandy was in France, and Gibraltar, a bit of rock next to Spain. Well, Brits lay claim to all these territories and we're not giving them back to anyone. After all, British Californias are much older and wiser, and we have more than our American cousins – more than seven, if you're counting.

*ambridgeshire... Falkirk... Norfolk... Suffolk... Dumfries & Galloway... Shropshire...*

The Brits have a long, sometimes troubling, history with gnomes. The father of the common garden gnome, Sir Charles Isham, became so besotted with the little fellers, originally used to protect miners, that he brought them back to Lamport Hall from Germany in the 19th century. Banned from the dining room table by Sir Charles' wife, he instead planted them in a 24-foot alpine rockery shrine, which he built in their honour. His only surviving adoptee, Lampy, a campaigner for gnome rights, is now allowed indoors. Today, gnomes have a new crusader, Ann Atkin, who also likes to gnome people. "I love turning people into gnomes," says Ann, "because seeing the world through a gnome's eyes makes everything more magical". At her Gnome Reserve & Wild Flower Garden in Devon, over 1000 gnomes go about their business, fishing, sunbathing, and peeking out from under the bushes. Visitors are loaned hats and fishing rods free of charge to make the locals feel more comfortable, while Ann paints pictures of her perfect world.

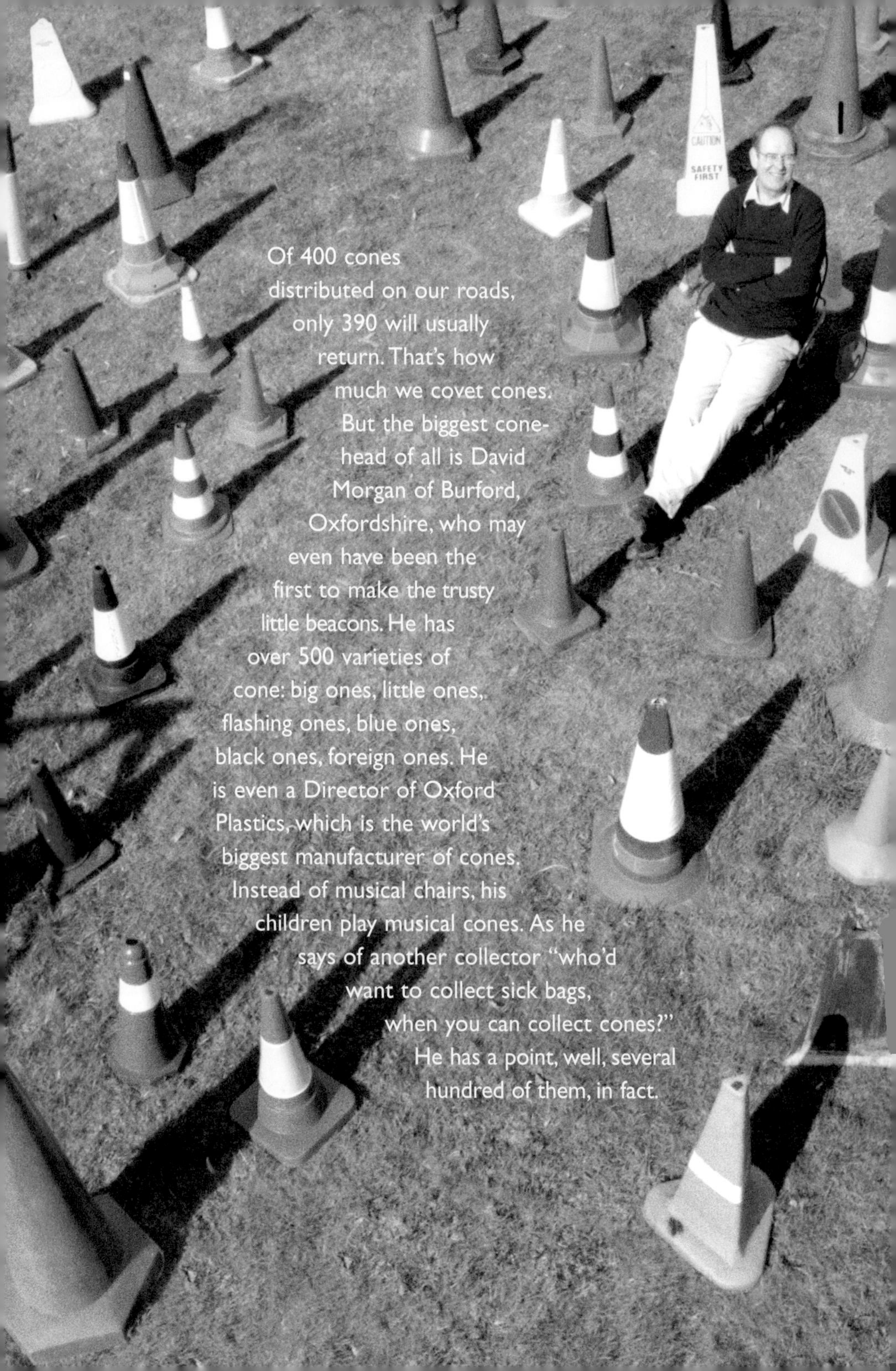


Of 400 cones distributed on our roads, only 390 will usually return. That's how much we covet cones. But the biggest cone-head of all is David Morgan of Burford, Oxfordshire, who may even have been the first to make the trusty little beacons. He has over 500 varieties of cone: big ones, little ones, flashing ones, blue ones, black ones, foreign ones. He is even a Director of Oxford Plastics, which is the world's biggest manufacturer of cones. Instead of musical chairs, his children play musical cones. As he says of another collector "who'd want to collect sick bags, when you can collect cones?" He has a point, well, several hundred of them, in fact.

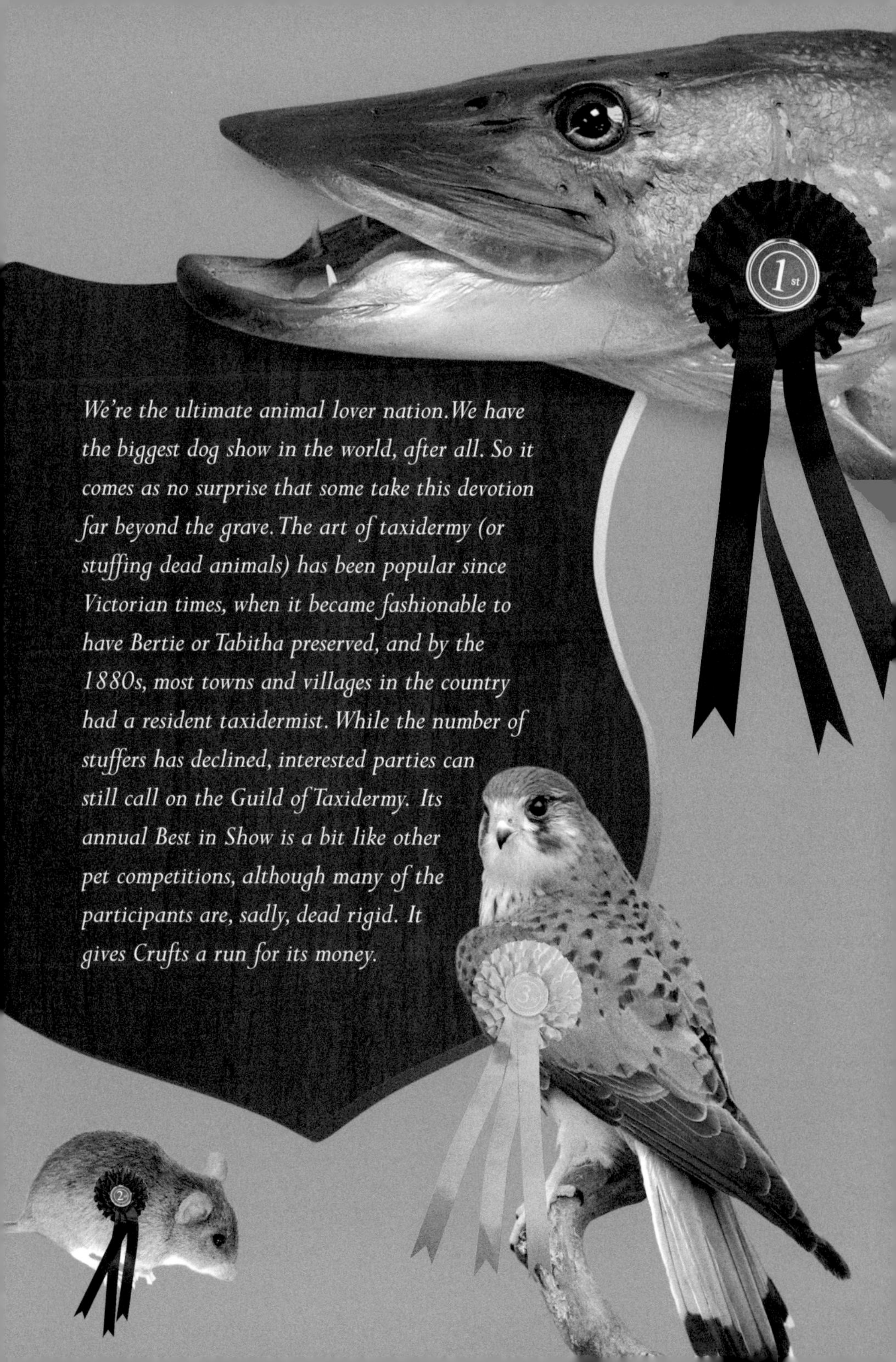

*We're the ultimate animal lover nation. We have the biggest dog show in the world, after all. So it comes as no surprise that some take this devotion far beyond the grave. The art of taxidermy (or stuffing dead animals) has been popular since Victorian times, when it became fashionable to have Bertie or Tabitha preserved, and by the 1880s, most towns and villages in the country had a resident taxidermist. While the number of stuffers has declined, interested parties can still call on the Guild of Taxidermy. Its annual Best in Show is a bit like other pet competitions, although many of the participants are, sadly, dead rigid. It gives Crufts a run for its money.*

The Ugley Women's Institute assembles once a month in Ugley's Ugley Village Hall, Essex.

# THERE'S NO ARM IN TOE WRESTLING

There's no arm at all in toe wrestling, but it can hurt. Five times World Champion Alan "Nasty" Nash (actually a very nice man) practises on nothing more than a broom handle. But competition is intense, with over 100 entrants taking part. To get through the knockout tournament, contestants must force over their opponent's foot in the best of three "toe-downs". Winners, having wrestled their way to victory, hobble their way to the Toedium. The sport was started when two locals at Ye Olde Royal Oak Inn, near Wetton, Derbyshire, settled on a world championship that the Brits, with their mighty toeholds, could always win. True to form, no fancy-foot-working foreigner has yet won the title.

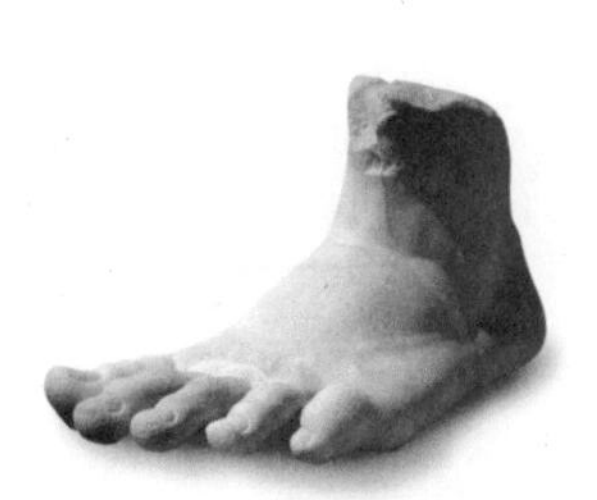

W Y N G Y L L G O G E R Y C H

Try getting your tongue around that. The UK's longest place name is in Wales, and manages to get away with no fewer than four Ls and three GOs in a row. The town's name was changed as a publicity stunt in the 19th century to attract tourists when the railway came through. Even locals find this mouthful tricky, so the town is more commonly known as Llanfair PG. The shortest place name is Ae in Dumfries & Galloway, Scotland.

# ST MARY'S CHURCH
# IN THE HOLLOW
# OF THE WHITE HAZEL

# NEAR A RAPID WHIRLPOOL
# AND THE CHURCH OF
# ST TYSILIO NEAR THE RED CAVE

# 34 Scrabble pts

(that's what we make it, anyway)

YRNDROBWLLLLAN
2,0
BBLE

# WESTWARD HO

Westward Ho!, the only town in Britain with an exclamation mark, is named after Charles Kingsley's book of the same name. The Victorian public was so captivated by the swashbuckling yarn that the town was actually built and named after the book in 1863. Rudyard Kipling also wrote Stalky & Co about his schooldays here.

# SIX FEET UNDER

Go out in unique style with Vic Fearn & Co's special coffin selection. They've made a canal narrowboat, a skip, a six-foot sports bag, an egg, a guitar, a wine cork and corkscrew, and a glass-topped coffee table. At the moment one of their coffins is sidelining as a stereo cabinet, while a giant bell is waiting for its owner to become a dead ringer. Director of the company, David Crampton, wants to be buried in a giant dancing shoe.

We love our lawns but nowhere more so than in Liverpool, where we've unearthed Menlove Gardens North, Menlove Gardens South and Menlove Gardens West, L18. There is no East. It appears that nothing grows in the East.

Even back in the 1830s people were potty about their gardens. Lords and ladies paid men to live as Ornamental Hermits in their grounds to entertain and surprise their guests. One hermit managed four years.

"Bread, nah, you mean Uncle Fred!" "I say, may I have a loaf?"

HAM ½
SANDWICH 3

Ham, just up the road from Sandwich, Kent.

“Och, aye, arl av a bap.”
“A free barm cake... give over!”
“Where is that baton to?”
“Ay up me duck, I'll av a cob.”
“Allreet pet, giy us a stottie cyek.”

"I'll have another consonant please, Carol."

*Countdown's* Richard Whiteley is the honorary mayor of Wetwang, East Yorkshire.

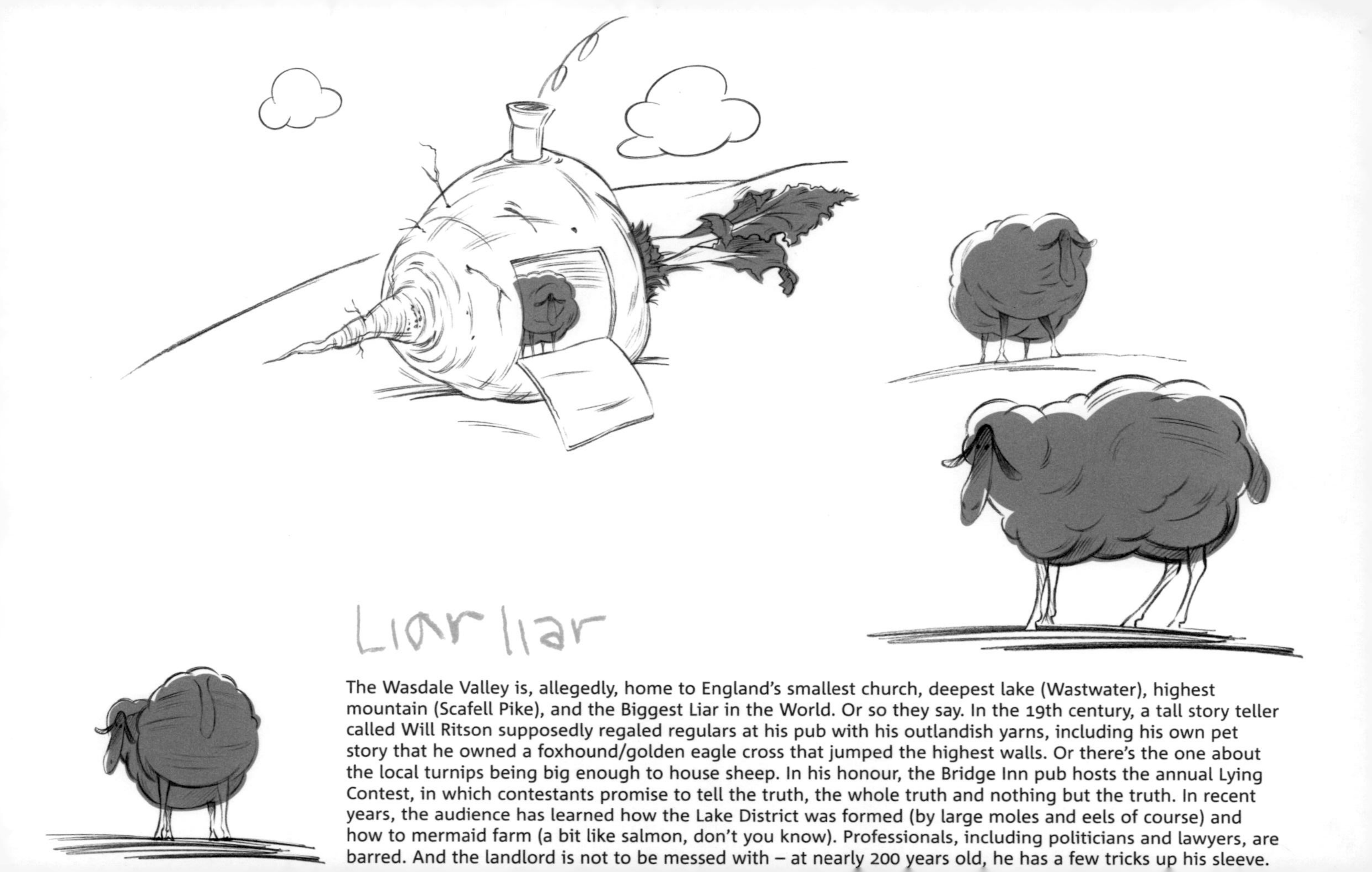

# Liar liar

The Wasdale Valley is, allegedly, home to England's smallest church, deepest lake (Wastwater), highest mountain (Scafell Pike), and the Biggest Liar in the World. Or so they say. In the 19th century, a tall story teller called Will Ritson supposedly regaled regulars at his pub with his outlandish yarns, including his own pet story that he owned a foxhound/golden eagle cross that jumped the highest walls. Or there's the one about the local turnips being big enough to house sheep. In his honour, the Bridge Inn pub hosts the annual Lying Contest, in which contestants promise to tell the truth, the whole truth and nothing but the truth. In recent years, the audience has learned how the Lake District was formed (by large moles and eels of course) and how to mermaid farm (a bit like salmon, don't you know). Professionals, including politicians and lawyers, are barred. And the landlord is not to be messed with – at nearly 200 years old, he has a few tricks up his sleeve.

# Leopard man

The Leopard man inhabits the Isle of Skye and can sometimes be spotted when he emerges from his natural habitat, the ruins of a remote croft. He may be scary to look at, but we are assured that he is quite harmless and can be safely approached. Tom happens to be the most tattooed man on the planet and sports a set of big cat fangs. He manages to survive quite happily in the wild, without electricity, water or central heating, although he can sometimes be seen in the local pub.

*Ladies fainted, dogs yelped and children fled in alarm (a young boy even broke his arm) when a strange man loomed large in a London street in 1797. The devilish chap who caused the riot was John Hetherington and he was promptly arrested for causing a breach of the peace and fined heavily. His crime? Well, he had "appeared on a public highway wearing a tall structure having a shining lustre, calculated to frighten timid people". As the details of the disturbance emerged, it appeared he was not intending to cause mayhem – but simply trying to model his new invention, now known as the Top Hat. Eventually Hetherington was vindicated when his hat became a fashion craze, after Prince Albert started wearing one in the 1850s.*

*The oldest hatters in London, Lock & Co, has been going for over 300 years, and has produced hats for the likes of Winston Churchill and Oscar Wilde. The company's files show that the British head has increased by three-eighths of an inch in circumference every half century. It is rumoured that Lewis Carroll took his inspiration for the Mad Hatter from the then Lock & Co proprietor, James Benning.*

## Allerton Mauleverer

Meaning: Farmstead of the Mauleverer family. First recorded: c. 12th century, but Allerton is in the Domesday Book. In: North Yorkshire.

## Praze-an-Beeble

Meaning: Cornish for meadow by the stream.
First recorded: 17th century. In: Cornwall.

## Stow Cum Quy

Meaning: Place with a cow island.
First recorded: c. 14th century. In: Cambridgeshire.

## Westonzoyland

Meaning: Enclosure in the Sowi land, from the River Sow. First recorded: 13th century. In: Somerset.

## Toller Porcorum

Meaning: On the river, Toller of the Pigs, named after the village's herd of swines.
First recorded: 14th century. In: Dorset.

## Ryme Intrinseca

Meaning: Ryme: Saxon for edge. Intrinseca: Latin for inner.
First recorded: Ryme mentioned in a survey in 1086, but only Latinised with Intrinseca in 15th century. In: Dorset.

## Compton Pauncefoot

Meaning: Named after a Norman knight called Pauncefote (meaning round-bellied) lying in a narrow valley. First recorded: In Norman times. In: Somerset.

*Huish* EPI

SCO

Meaning: Household of the Bishop. First recorded: c. 973 when it was Hiwissh. In: Somerse

PI

Subject to remaining records, lost facts and figures, and hearsay, like all of the best history.

# IDLE

## WORKING MEN'S CLUB

YORKSHIRE

The small town of Idle, near Bradford, is home to many an Idle person, and the site of the Idle Working Men's Club.

Any Idle person around the world is welcome to join; in fact, even Idle ladies can put their feet up and become members.

With all the Idle talk that must go on, who is left to pull the pints?

# WORK SLOG PLAY

Brits go a bit barmy at the seaside – that's what happens when all that pent-up tension is released from the hardest workers in Europe. Each year we'll consume ten million sticks of rock, over one million ice cream cones, 500,000 candy flosses, 600 kilos of gobstoppers...

*...and that's just in Blackpool!*

KISS ME
QUICK

Even donkeys are let loose here. It's all laid out in the 1942 Donkey's Charter, which ensures they get at least one day off a week: "No asses shall stand, or ply for work on the sands on any Friday". But equally, there should be no funny business near the beach: "No stud shall ply for hire within 50 yards of the hulking or sea wall". And who'd want to get a name for being a bad ass.

Ageing donkeys can now retire to Devon's Donkey Sanctuary, but they may prefer to graze in Surrey's Donkey Town.

# "Come on Cecil!"

The annual snail-bitinc

contenders slugging it out on a rigorous 13-inch stretch of racinc

course. Punters can find it hard to make up their minds about which

hide in their shells." The most hotly contested event in the racinc

Over 200 trainers come from all over the country to put gastropods

themselves on their area's superior breeding conditions. As Hilary

racing what Newmarket is to horse racing. It's damp and low lying

The 13-inch world sprint record – currently held by Archie at two

Like all the best thoroughbreds, the winner goes out to stud.

spectacular of the British Snail Racing Association sees

track, some taking as little as a few minutes to finish the

snails to back. "It's hard to study the form when they

calendar is the Snail Racing World Championships.

through their paces at Congham, but locals pride

Scase, organiser of the event, says: "Congham is to snail

and so a very good area for breeding fast snails."

minutes precisely – has yet to be broken.

# sticky pints

Is it the wallpaper glue, the tobacco and the grime? Or is it just pure magic? Regulars at the Somerset House in Stourbridge don't just rest their pints on the table – they can also stick them to the walls, and they can dangle there for hours on end. Careful about brushing your sleeve against the wall as you may end up hanging out there longer than you intended.

Booze is a hamlet in North Yorkshire. The Red Lion pub, Langthwaite, of All Creatures Great and Small fame, is in stumbling distance.

Pick your poison and stick it to the wall.

## LEICESTERSHIRE ANGLER USING JELLY BABY BAIT FINDS THAT COD FAVOUR BLACK ONES, EEL CRAVE ORANGE, WHILE CATFISH PREFER PINK

It's all a bit fishy, if some UK place names are anything to go by. Trouts is in Kent, while Dolphin is in Flintshire. Then there's Eel Pie Island in Richmond, and the Angler's Retreat in Ceredigion. But when all's said and done, The Fish is in Shropshire.

# Six Mile Bottom

Six Mile Bottom is six miles from Newmarket, and in the bottom of a valley, hence the cheeky name. The local pub, Green Man, was used as a staging post from the 15th century and King Charles II came here to frequent his famous mistress, Nell Gwynn. Her ghost is said to still wander the corridors...

Hundreds of UK place names are
inspired by our favourite preoccupation.

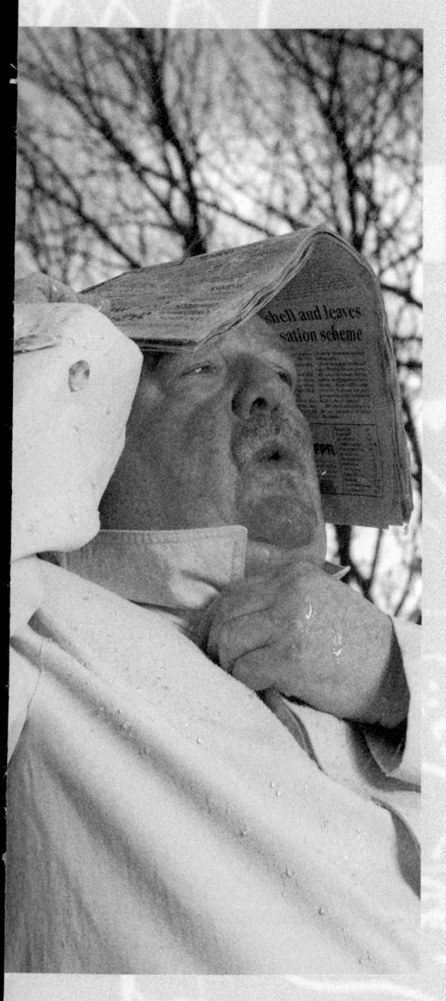


**Bright and breezy in Quaking Houses, with a chance of showers in Windy Walls, turning to sleet and possible snow in Sunset, clearing with a frosty start to the morning in Snow End, followed by persistent mist in Clouds, and an icy cold, southwesterly gale in Fog Close, resulting in a serious weather warning on Rainbow Hill.**

**So not too bad overall.**

**And that's only half of it...**

We haven't even had time to tell you about the Nasty people in Herts, or Slop Bog Nature Reserve (actually quite picturesque). There's Bedlam in North Yorkshire, Motorcycle Funerals, Bog Snorkelling, Pearly Kings and Queens, Captain Beany and Devon's Dog Village. The Maldon Mud Race, Swine in Yorkshire, the British Federation of Ferret Welfare, Whip Ma Whop Ma Gate in York and the Naked Rambler, no doubt avoiding Nut Crackers in Devon. The people living in the numerous places called Nomansland. The World Black Pudding Throwing Championships, Conundrum in Northumberland, the Extreme Ironing Bureau (ironing up mountains, in rivers and underwater), the World French Knitting Champion Ted Hannaford, Nether Wallop, and Dirty Gutter. Ashby-de-la-Zouch, the Leg of Mutton and Cauliflower pub, nettle eating competitions, the Hunting of the Earl of Rhone, and Professor Len Fisher, pioneer of the Physics of Biscuit Dunking. Haggis Hurling, the London Vampyre (sic) Group, fried chocolate sandwiches, Garry Stretch Turner (who attached 153 clothes pegs to his face) and his Circus of Horrors, the Crooked House pub, Druids, Princess Caribou, Yelling in Cambridgeshire, and Captain Cutlass, the plank-walking champion. Gurning, the Lawnmower Racing Society, the Bucket of Blood pub, the fearful scarecrow competitions, Peveril of the Peak, the real Fawlty Towers, Basil! The British Toilet Association, the Burry Man, Louis Tussauds Waxworks, the Double Decker B&B, Cheese Rolling. Colin Boness the 45mph Trolley Racer, Wincle, the Peter Pan Cup, Wincanton, twinned with Pratchett's made-up Ankh-Morpork. The Fairy Ring, the worm charming championship, Denby Dale's 12 tonne giant pie, Wallish Walls... It's never Dull in Britain, except in the village in Perth & Kinross.